CLIMATE CHANGE &
HUMAN

Detailed information on climate change

Wanda V Amaro

Copyright

Table of Content

Climate Change & Politics

Climate Change & Food Security

Climate Change & Technological Innovations

Climate Change Denial & Misinformation

Climate Change & Public Awareness

Climate Change

Climate change is a significant and pressing issue that has emerged as one of the most significant challenges facing humanity today. The term "climate change" refers to the long-term changes in the earth's climate, including rising temperatures, sea levels, and changes in weather patterns caused by increased concentrations of greenhouse gasses in the atmosphere.

The scientific consensus is that the planet is warming at an unprecedented rate, with global temperatures rising by approximately 1.1 degrees Celsius since the pre-industrial era. This warming trend is primarily driven by the burning of fossil fuels, deforestation, and other human activities that release carbon dioxide and other greenhouse gasses into the atmosphere.

The consequences of climate change are already being felt around the world, with more frequent and severe natural disasters, such as hurricanes, droughts, and wildfires, becoming commonplace. Rising sea levels are also threatening coastal communities and island nations, with low-lying areas at risk of flooding and displacement. Climate change is also causing significant ecological disruptions, with the loss of biodiversity and the extinction of many species.

Despite the severity of the situation, there are solutions that can help mitigate the worst impacts of climate change. One of the most effective strategies is to transition to renewable energy sources such as wind, solar, and geothermal power, which produce little to no greenhouse gas emissions. Other solutions include improving energy efficiency in buildings and appliances, reducing food waste and consumption, and investing in sustainable transportation options.

Addressing climate change also requires collective action and international cooperation. The Paris

Agreement, a landmark international treaty signed in 2015, seeks to limit global warming to well below 2 degrees Celsius, with an aim to limit the increase to 1.5 degrees Celsius. Achieving these goals will require significant efforts from governments, businesses, and individuals around the world.

Climate change is an urgent and complex issue that requires immediate action. It is a threat to our planet's ecosystems, economies, and societies, and the consequences of inaction will be severe. However, by adopting sustainable practices, investing in renewable energy, and working collaboratively to reduce greenhouse gas emissions, we can mitigate the worst impacts of climate change and create a more sustainable future.

Greenhouse Gas Emissions

Greenhouse gas emissions are one of the primary drivers of climate change. These gasses, including carbon dioxide (CO2), methane (CH4), and nitrous oxide (N2O), trap heat in the atmosphere and contribute to the greenhouse effect, which is the warming of the earth's surface.

Human activities are the primary source of greenhouse gas emissions, with the burning of fossil fuels for energy production and transportation being the largest contributor. Other human activities that contribute to greenhouse gas emissions include deforestation, agriculture, and industrial processes.

Carbon dioxide, the most abundant greenhouse gas, is emitted by the burning of fossil fuels such as coal, oil, and natural gas. These fuels release carbon dioxide into the atmosphere when they are burned, and the concentration of CO2 in the atmosphere has increased by 45% since the Industrial Revolution. The majority of CO2 emissions come from the energy sector, with transportation being the largest contributor.

Methane, a potent greenhouse gas, is emitted from a variety of sources, including livestock production, rice cultivation, and natural gas production and distribution. Methane is a more potent greenhouse gas than CO2, but its concentration in the atmosphere is much lower. However, reducing methane emissions can be an effective way to slow down the rate of warming.

Nitrous oxide is another potent greenhouse gas that is emitted from agricultural practices, such as the use of nitrogen fertilizers, and industrial processes such as the production of nitric acid.

The impact of greenhouse gas emissions on the climate system is significant, with global temperatures increasing by approximately 1.1

degrees Celsius since the pre-industrial era. This warming trend is causing significant impacts, including rising sea levels, more frequent and severe natural disasters, and the loss of biodiversity.

To address the issue of greenhouse gas emissions, it is important to reduce the amount of greenhouse gasses released into the atmosphere. This can be achieved through a variety of measures, including investing in renewable energy sources, improving energy efficiency, reducing food waste and consumption, and transitioning to low-emission transportation options.

Greenhouse gas emissions are a major contributor to climate change, and their impact on the planet is significant. Reducing greenhouse gas emissions is critical to mitigating the worst impacts of climate change, and this requires collective action from governments, businesses, and individuals around the world. By taking steps to reduce greenhouse gas emissions, we can create a more sustainable future for ourselves and future generations.

Rising Sea Levels

Rising sea levels are one of the most significant consequences of climate change, and they pose a significant threat to coastal communities around the world. The warming of the planet is causing the polar ice caps and glaciers to melt, leading to a rise in sea levels. The sea level has already risen by approximately 20 centimeters since the beginning of the 20th century, and it is expected to rise by another 30-60 centimeters by the end of this century, depending on the extent of greenhouse gas emissions.

The impact of rising sea levels on coastal communities is significant and multifaceted. One of the most direct effects is coastal erosion, which occurs when waves and tides wear away the coastline. Coastal erosion can cause property damage, loss of habitat, and even force communities to relocate.

Another impact of rising sea levels is an increase in the frequency and severity of coastal flooding. Even small increases in sea level can lead to more frequent and severe coastal flooding, which can damage infrastructure and homes, threaten human health, and lead to economic losses.

In addition, rising sea levels can exacerbate the impact of storms and natural disasters, such as hurricanes and tsunamis. As sea levels rise, storm surges become more severe, causing more extensive flooding and damage to coastal communities.

The impact of rising sea levels is not limited to physical damage. Coastal communities also face significant social and economic challenges. Many communities rely on tourism and fishing, which can be disrupted by coastal erosion, flooding, and other impacts of rising sea levels. Additionally, low-income and marginalized communities are often disproportionately affected by rising sea levels, as they may lack the resources to relocate or rebuild after a disaster.

To mitigate the impact of rising sea levels on coastal communities, it is important to take a multi-faceted approach. This can include measures such as improving coastal infrastructure to reduce the risk of flooding, creating coastal buffers such as wetlands and sand dunes, and developing early warning systems for storms and natural disasters. In addition, communities may need to consider relocation to safer areas in the long-term, particularly in areas where the risk of coastal erosion and flooding is high.

Rising sea levels pose a significant threat to coastal communities, with impacts ranging from physical damage and erosion to social and economic disruption. Mitigating the impact of rising sea levels requires a comprehensive approach that addresses both the immediate and long-term impacts of climate change. By taking action to reduce greenhouse gas emissions and adapt to the impacts of rising sea levels, we can create more resilient and sustainable coastal communities.

Climate Change Impact on Biodiversity & Wildlife

Climate change is having a significant impact on biodiversity and wildlife around the world. The Earth's temperature is rising at an unprecedented rate, which is leading to changes in weather patterns, rising sea levels, and the loss of habitats. These changes are having a severe impact on biodiversity, which is the variety of life on Earth, and on the ecosystems that support it. In this article, we will explore the effects of climate change on biodiversity and wildlife, the causes of these effects, and what can be done to mitigate them.

Effects of Climate Change on Biodiversity and Wildlife

Climate change is causing significant changes to the habitats of wildlife species around the world. Many animals and plants are struggling to adapt to these changes, and some species are facing extinction. Rising temperatures are causing changes in the timing of seasonal events, such as flowering and migration patterns, which can lead to mismatches in food availability and reproductive cycles. This, in turn, can affect the survival and reproduction of species.

The warming of the planet is also causing the melting of glaciers and sea ice, leading to rising sea levels. This is particularly problematic for coastal habitats, which are home to a large number of species. As sea levels rise, many of these habitats are becoming flooded, which can lead to the loss of important breeding grounds and feeding areas.

Changes in precipitation patterns are also affecting biodiversity. Droughts are becoming more frequent and severe in some regions, which can lead to the loss of vegetation and habitat for wildlife. Heavy rainfall events are also becoming more common, leading to flooding and the destruction of habitats.

The impacts of climate change are not limited to individual species or habitats. The loss of biodiversity can have far-reaching consequences for entire ecosystems, which can affect the provision of ecosystem services such as pollination, nutrient cycling, and climate regulation. This can, in turn, affect human well-being, as many of our basic needs, such as food and clean water, rely on healthy ecosystems.

Causes of Climate Change on Biodiversity and Wildlife

The primary cause of climate change is the release of greenhouse gasses, such as carbon dioxide and methane, into the atmosphere. These gasses trap heat from the sun and cause the Earth's temperature to rise. The main sources of these gasses are human activities, such as burning fossil fuels, deforestation, and industrial processes.

As the world's population continues to grow, the demand for resources is increasing, which is driving up greenhouse gas emissions. The expansion of agriculture and urban areas is also leading to the loss of natural habitats, which is contributing to the loss of biodiversity.

What can be done to Mitigate the Impact of Climate Change on Biodiversity and Wildlife?

To mitigate the impact of climate change on biodiversity and wildlife, it is essential to reduce greenhouse gas emissions. This can be achieved through a combination of measures, including reducing energy consumption, increasing the use of renewable energy sources, and improving energy efficiency in buildings and transportation.

Protecting and restoring natural habitats is also crucial for the survival of many wildlife species. This includes protecting areas of high biodiversity, such as tropical rainforests, and restoring degraded habitats, such as wetlands and grasslands. It is also essential to reduce the impact of human activities on wildlife, such as overfishing and poaching.

Another important strategy for mitigating the impact of climate change on biodiversity is to promote adaptation. This includes supporting the development of strategies that help wildlife species and ecosystems adapt to the changing climate. For example, restoring degraded habitats can provide refuge for species that are struggling to adapt to changing conditions.

Climate change is having a significant impact on biodiversity and wildlife around the world. Rising temperatures, changing precipitation patterns, and rising sea levels are leading to the loss of habitats and the extinction of species. To mitigate these impacts, it is essential to reduce greenhouse gasses.

Climate Change and Extreme Weather Events

Climate change has been identified as a major contributor to the increase in extreme weather events such as hurricanes, floods, and droughts. As the Earth's temperature continues to rise due to human activities, the frequency and intensity of these events are expected to increase, leading to severe consequences for human societies and ecosystems around the world. In this article, we will explore the relationship between climate change and extreme weather events and the scientific evidence supporting this connection.

The Link between Climate Change and Extreme Weather Events

The scientific community has long recognized the relationship between climate change and extreme weather events. The Intergovernmental Panel on Climate Change (IPCC) has stated that "there is high confidence that the frequency and intensity of some extreme weather events have increased in recent decades, and that human-caused climate change is a contributing factor."

Extreme weather events such as hurricanes, floods, and droughts are primarily caused by changes in atmospheric conditions, such as temperature, humidity, and wind patterns. Climate change affects these conditions in several ways, leading to an increase in the frequency and intensity of extreme weather events.

For example, warmer air and ocean temperatures can lead to stronger and more frequent hurricanes. Rising sea levels can increase the impact of storm surges and flooding associated with hurricanes. Similarly, changes in precipitation patterns due to climate change can lead to more frequent and severe floods and droughts.

Scientific Evidence Supporting the Link

The link between climate change and extreme weather events is supported by a vast body of scientific evidence. For example, a study published in the journal Nature in 2017 found that climate change had increased the likelihood of the severe rainfall associated with Hurricane Harvey, which caused extensive flooding in Houston, Texas in 2017.

Similarly, a study published in the journal Science in 2018 found that climate change had increased the likelihood of the drought conditions that led to the severe wildfires that ravaged California in 2017. Other studies have linked climate change to an increase in the frequency and intensity of heat waves, which can have severe impacts on human health and infrastructure.

Mitigating the Impacts of Climate Change on Extreme Weather Events

To mitigate the impacts of climate change on extreme weather events, it is essential to reduce greenhouse gas emissions, which are the primary cause of climate change. This can be achieved through a combination of measures, including reducing energy consumption, increasing the use of renewable energy sources, and improving energy efficiency in buildings and transportation.

In addition, adaptation measures can help communities prepare for and respond to the impacts of extreme weather events. This can include developing early warning systems, improving infrastructure resilience, and implementing land-use planning measures that reduce vulnerability to floods and droughts.

The link between climate change and extreme weather events such as hurricanes, floods, and droughts is well established by scientific evidence. As the Earth's temperature continues to rise due to human activities, the frequency and intensity of these events are expected to increase, leading to severe consequences for human societies and ecosystems around the world. To mitigate these impacts, it is essential to reduce greenhouse gas emissions and implement adaptation measures that help communities prepare for and respond to extreme weather events.

The Melting of Polar Ice Caps

The melting of polar ice caps, particularly those in the Arctic and Antarctic, has been a major concern for scientists and policymakers for decades. As temperatures continue to rise due to human-induced climate change, the rate of ice melting has accelerated, leading to rising sea levels that threaten to have significant impacts on human populations and the environment.

To understand the impact of melting polar ice caps on global sea levels, it's important to first understand the role that these ice caps play in the Earth's climate system. The polar ice caps act as "reflectors," reflecting a large amount of incoming solar radiation back into space and helping to regulate the Earth's temperature. This is due to the high albedo, or reflectivity, of ice and snow compared to other surfaces.

As temperatures warm, however, the reflective surface of the ice caps begins to melt and expose darker, less reflective surfaces beneath. This leads to a positive feedback loop, where more solar radiation is absorbed, causing further melting and thus more absorption of solar radiation. This feedback loop can lead to a rapid acceleration of ice melting and contribute to further warming of the Earth's climate.

One of the primary impacts of melting polar ice caps is rising sea levels. The amount of ice locked up in polar ice caps is enormous – the Greenland ice sheet alone contains enough ice to raise global sea levels by 7.2 meters (23.6 feet) if it were to melt entirely. The Antarctic ice sheet contains even more ice, and its melting could raise sea levels by an additional 58 meters (190 feet).

While it's unlikely that the polar ice caps will melt completely in the near future, even a small amount of melting can have significant impacts on sea levels. According to the Intergovernmental Panel on Climate Change (IPCC), global sea levels rose by around 15 cm (6 inches) during the 20th century, and are projected to rise by an additional 26 to 82 cm (10 to 32 inches) by the end of the 21st century. However, some recent studies suggest that these projections may be conservative, and that sea levels could rise much more rapidly if melting of the polar ice caps continues at its current rate.

The impacts of rising sea levels can be severe. Coastal cities and low-lying areas are particularly vulnerable to flooding and storm surges, which can cause billions of dollars in damage and displace millions of people. In addition to the direct impacts on human populations, rising sea levels can also have significant ecological consequences, including the loss of critical habitats such as mangrove forests and coral reefs.

To address the melting of polar ice caps and its impacts on sea levels, it's important to take aggressive action to reduce greenhouse gas emissions and limit global warming. This will require significant policy changes, including transitioning to renewable energy sources, improving energy efficiency, and implementing policies to reduce emissions from transportation and industry. Additionally, efforts to adapt to rising sea levels may include building sea walls and other infrastructure to protect vulnerable coastal areas, as well as developing new technologies to help mitigate the impacts of climate change.

The melting of polar ice caps is a major concern for the Earth's climate system, with significant impacts on global sea levels and human populations. Addressing this issue will require concerted action to reduce greenhouse gas emissions and adapt to the impacts of climate change.

Renewable Energy

Renewable energy sources such as solar, wind, hydro, geothermal, and biomass are increasingly being seen as critical to reducing carbon emissions and combating climate change. Unlike traditional fossil fuels, renewable energy sources generate electricity without producing greenhouse gasses, which makes them a promising alternative to mitigate the negative impacts of climate change.

The potential of renewable energy sources to reduce carbon emissions is significant. According to the International Energy Agency (IEA), renewable energy sources accounted for over 70% of global net additions to power generation capacity in 2019, with solar and wind power leading the way. This trend is expected to continue, with the IEA projecting that renewable energy sources will provide nearly 90% of the increase in global power capacity between 2020 and 2025.

One of the key benefits of renewable energy sources is their ability to replace traditional fossil fuels in the generation of electricity. Fossil fuels such as coal, oil, and natural gas are the primary sources of carbon emissions, accounting for over two-thirds of global greenhouse gas emissions. By transitioning to renewable energy sources, we can reduce our reliance on these fossil fuels and significantly reduce carbon emissions.

Solar and wind power are two of the most promising renewable energy sources for reducing carbon emissions. Solar energy is abundant, free, and available almost everywhere on the planet. Advances in solar panel technology have made it more efficient and affordable, and it is now one of the fastest-growing energy sources in the world. Wind energy is similarly abundant and has the potential to provide a significant portion of the world's electricity needs. As wind turbine technology improves, the cost of generating wind power is expected to continue to decrease, making it increasingly competitive with traditional fossil fuels.

In addition to solar and wind power, other renewable energy sources such as hydroelectricity, geothermal, and biomass can also play a role in reducing carbon emissions. Hydroelectricity is a well-established technology that has been in use for over a century. It generates electricity by harnessing the power of moving water, and it is one of the most efficient and reliable renewable energy sources. Geothermal energy uses the Earth's natural heat to generate electricity, and it has the potential to provide a significant amount of power in regions with high geothermal activity. Biomass energy, which involves the conversion of organic materials into energy, is another promising renewable energy source that can help reduce carbon emissions.

The potential of renewable energy sources to reduce carbon emissions is not limited to electricity generation. Transportation is another major source of carbon emissions, accounting for over one-quarter of global greenhouse gas emissions. The use of electric vehicles (EVs) powered by renewable energy sources such as solar and wind power can significantly reduce the carbon footprint of transportation. Additionally, the production of biofuels from renewable sources such as corn, sugarcane, and switchgrass can also reduce carbon emissions from transportation.

Renewable energy sources have the potential to significantly reduce carbon emissions and mitigate the negative impacts of climate change. While there are still challenges to be addressed, such as intermittency and storage, the rapid growth of renewable energy sources in recent years is a positive sign of the potential for a sustainable future. Policymakers, businesses, and individuals all have a role to play in accelerating the transition to renewable energy sources and reducing our reliance on fossil fuels.

Deforestation and land use change

Deforestation and land use change are significant contributors to climate change, accounting for up to 20% of global greenhouse gas emissions. The destruction of forests and the conversion of land for agriculture, pasture, and urbanization release carbon into the atmosphere and disrupt the Earth's natural carbon cycle.

Forests play a critical role in regulating the Earth's climate. Trees absorb carbon dioxide from the atmosphere during photosynthesis and store it in their biomass and soil. When forests are destroyed or degraded, this carbon is released back into the atmosphere, contributing to climate change. Deforestation is responsible for approximately 10% of global greenhouse gas emissions, making it one of the largest sources of emissions after the burning of fossil fuels.

In addition to releasing carbon, deforestation also contributes to climate change through the loss of the many benefits that forests provide. Forests act as carbon sinks, absorbing carbon dioxide from the atmosphere and storing it in trees and soil. They also regulate the water cycle, provide habitat for wildlife, and protect soil from erosion. The loss of forests can lead to soil degradation, loss of biodiversity, and increased frequency and intensity of natural disasters such as floods and landslides.

Land use change, which includes the conversion of forested land to agriculture, pasture, or urbanization, is another major contributor to climate change. The clearing of land for these purposes releases carbon into the atmosphere and can result in soil degradation, loss of biodiversity, and changes in the local climate. The expansion of agriculture and pastureland is a significant driver of land use change, with large areas of tropical forest being cleared for cattle ranching and soybean production.

The impacts of deforestation and land use change on climate change are not limited to their direct effects on carbon emissions. These activities also have indirect impacts on the climate through changes in the Earth's albedo, which is the amount of sunlight reflected back into space. Forests have a high albedo, reflecting more sunlight than agricultural land or urban areas. When forests are cleared, the darker surface of the ground absorbs more sunlight, contributing to local warming and altering regional and global weather patterns.

Reducing deforestation and land use change is critical to mitigating the impacts of climate change. This can be achieved through a combination of policies, incentives, and technological solutions. Policies such as forest conservation and restoration programs, protected area designations, and carbon pricing can help to reduce deforestation and promote the sustainable management of forests. Incentives such as payments for ecosystem services and sustainable land use practices can also encourage farmers and landowners to adopt more sustainable practices.

Technological solutions such as satellite monitoring, remote sensing, and machine learning can also play a role in reducing deforestation and land use change. These technologies can provide real-time information on forest loss and land use change, enabling policymakers and landowners to take action to prevent further degradation. Additionally, sustainable land management practices such as agroforestry, which combines tree planting with agriculture, can help to reduce the impacts of land use change and improve soil health.

Deforestation and land use change are significant contributors to climate change, releasing carbon into the atmosphere and disrupting the Earth's natural carbon cycle. Reducing deforestation and promoting sustainable land use practices are critical to mitigating the impacts of climate change and achieving a sustainable future.

The Economic Cost Of Climate Change

Climate change is one of the greatest challenges facing the world today, with potentially devastating economic consequences. As temperatures rise, extreme weather events such as floods, droughts, and heatwaves are becoming more frequent and intense, resulting in significant economic costs.

The economic costs of climate change are multifaceted and can affect various sectors of the economy. For instance, the agricultural sector is vulnerable to climate change impacts such as changes in temperature and rainfall patterns, leading to crop failure and decreased productivity. Additionally, the increased frequency and intensity of natural disasters such as hurricanes, floods, and wildfires can lead to significant damage to infrastructure and property, resulting in high costs of repairs and rebuilding.

The costs of climate change can also manifest in the form of health impacts such as heat stress, respiratory illnesses, and vector-borne diseases, which can lead to increased healthcare costs and decreased workforce productivity.

However, the shift towards a low-carbon economy presents an opportunity for the creation of green jobs that can contribute to sustainable economic growth while mitigating the negative impacts of climate change. Green jobs are those that are environmentally friendly and contribute to the reduction of greenhouse gas emissions, such as jobs in renewable energy, energy efficiency, and sustainable transportation.

According to the International Labour Organization (ILO), the transition to a green economy has the potential to create 24 million new jobs globally by 2030, while simultaneously reducing greenhouse gas emissions by up to 36 percent.

Moreover, the transition to a green economy can also lead to cost savings and increased competitiveness, as companies that adopt sustainable practices can reduce their energy costs, improve resource efficiency, and enhance their brand image, thereby attracting more customers and investors.

The economic costs of climate change are significant, but the transition towards a green economy presents an opportunity for sustainable economic growth and the creation of green jobs. By investing in renewable energy, energy efficiency, and sustainable transportation, countries can mitigate the negative impacts of climate change while simultaneously contributing to sustainable economic development.

Climate change adaptation strategies

Climate change is a significant challenge for vulnerable communities, as they often lack the resources and infrastructure needed to adapt to its impacts. However, there are several climate change adaptation strategies that can help such communities to better cope with the effects of climate change.

One such strategy is the development of early warning systems to alert communities to impending extreme weather events, such as floods or storms. These systems can help to minimize the damage caused by these events and provide people with time to prepare and evacuate if necessary.

Another strategy is the implementation of climate-resilient infrastructure, such as sea walls, flood barriers, and drought-resistant crops, which can help to minimize the damage caused by extreme weather events and protect vulnerable communities from their impacts.

Furthermore, community-based adaptation approaches that involve local communities in decision-making and planning can be particularly effective in ensuring that adaptation measures are appropriate and relevant to the needs of the community. This approach can also help to build resilience and enhance social cohesion within communities.

In addition, promoting sustainable livelihoods, such as through agroforestry, can help communities adapt to climate change by providing alternative sources of income and reducing their dependence on vulnerable livelihoods that are sensitive to climate variability.

Finally, improving access to climate finance can help to support the implementation of climate change adaptation measures in vulnerable communities, particularly those in developing countries that often lack the resources to implement such measures themselves.

Climate change adaptation strategies for vulnerable communities should prioritize early warning systems, climate-resilient infrastructure, community-based adaptation, sustainable livelihoods, and improved access to climate finance. By implementing these strategies, vulnerable communities can better cope with the impacts of climate change and enhance their resilience to future climate variability.

Climate Change & Justice

Climate justice is an important concept that emphasizes the ethical and moral dimensions of climate change, particularly in relation to its disproportionate impact on marginalized groups. Climate change affects people and communities in different ways, and marginalized groups such as low-income communities, indigenous peoples, and people of color often face the most severe impacts.

For example, low-income communities are more likely to live in areas that are vulnerable to the impacts of climate change, such as flood-prone areas or areas with poor air quality, which can lead to increased health risks. Indigenous peoples, who rely heavily on natural resources for their livelihoods, often experience the loss of traditional lands and resources due to changes in weather patterns and environmental degradation.

Furthermore, people of color often face higher levels of exposure to environmental pollutants and hazardous materials due to systemic inequalities and environmental racism, which can lead to increased health risks and decreased quality of life.

Climate justice seeks to address these disproportionate impacts of climate change on marginalized groups by promoting equitable and just responses to climate change. This includes ensuring that the voices and perspectives of marginalized groups are heard and considered in decision-making processes related to climate change.

Additionally, climate justice involves addressing the root causes of climate change, including systemic inequalities and environmental degradation. This includes promoting policies and practices that prioritize the needs and interests of marginalized groups, such as investing in renewable energy, promoting sustainable agriculture, and supporting community-led adaptation and mitigation efforts.

Climate justice is an important concept that highlights the need to address the disproportionate impact of climate change on marginalized groups. By promoting equitable and just responses to climate change, we can ensure that the most vulnerable communities are protected and supported in the face of this global challenge.

Climate Change & Politics

The politics of climate change and international climate agreements have become increasingly important in recent years as the world seeks to address the challenge of global warming. Climate change is a global problem that requires global solutions, and international cooperation is essential to address this challenge effectively.

International climate agreements are critical to achieving this cooperation, as they provide a framework for countries to work together to reduce greenhouse gas emissions and mitigate the impacts of climate change. The most significant international climate agreement to date is the Paris Agreement, which was adopted in 2015 by 195 countries.

The Paris Agreement sets a goal of limiting global temperature rise to well below 2 degrees Celsius above pre-industrial levels, with a target of pursuing efforts to limit the temperature increase to 1.5 degrees Celsius. Countries are required to submit nationally determined contributions (NDCs), outlining their plans to reduce greenhouse gas emissions and adapt to the impacts of climate change.

However, the politics of climate change can complicate efforts to reach international agreements. Some countries, particularly those heavily reliant on fossil fuels, may resist efforts to reduce emissions or may argue that the burden of reducing emissions should be shared more equitably among countries.

Furthermore, there are concerns about the implementation and enforcement of international climate agreements. The Paris Agreement, for example, relies on voluntary commitments from countries, and there is no formal mechanism to enforce compliance with these commitments.

Despite these challenges, international climate agreements remain essential to addressing the global challenge of climate change. They provide a framework for countries to work together, set goals and targets, and establish mechanisms for monitoring progress and reviewing commitments.

Moreover, international climate agreements have the potential to promote innovation and investment in clean energy and sustainable practices, driving economic growth and job creation while also reducing greenhouse gas emissions.

The politics of climate change and international climate agreements are complex and challenging, but they are essential to address the global challenge of climate change effectively. Through international cooperation and commitment, we can work together to mitigate the impacts of climate change and build a more sustainable and resilient future for all.

Climate Change & Food Security

Climate change is having a significant impact on agriculture and food security, with potential consequences for global food systems and the livelihoods of millions of people. Rising temperatures, changing rainfall patterns, and more frequent extreme weather events are all contributing to these impacts.

One of the most significant impacts of climate change on agriculture is the reduction in crop yields. Rising temperatures and changing rainfall patterns can lead to droughts, floods, and other weather-related disasters that can damage crops and reduce yields. This can lead to food shortages and higher food prices, particularly in areas that are already struggling with food insecurity.

Moreover, climate change is affecting the quality and nutritional value of crops. For example, rising temperatures can reduce the protein content of crops, making them less nutritious for people and animals. Changes in rainfall patterns can also affect soil quality, making it harder to grow healthy crops.

Climate change is also affecting livestock production, as animals are vulnerable to extreme weather events such as heatwaves and floods. This can lead to decreased productivity and increased mortality rates, reducing the availability of meat and dairy products.

The impacts of climate change on agriculture and food security are particularly acute in developing countries, where agriculture is a significant source of livelihoods and food insecurity is already a significant issue. Small-scale farmers in these countries are particularly vulnerable to the impacts of climate change, as they often lack the resources and infrastructure needed to adapt to changing weather patterns.

To address these challenges, it is essential to develop adaptation strategies that promote climate-resilient agriculture and protect vulnerable communities from the impacts of climate change. This includes promoting sustainable agriculture practices that can improve soil health, increase crop yields, and reduce greenhouse gas emissions.

In addition, promoting the use of climate-smart technologies, such as drought-resistant crops, can help to build resilience and protect crops from the impacts of climate change. Supporting small-scale farmers and improving their access to credit, training, and markets can also help to strengthen their resilience to climate change and improve food security.

The impact of climate change on agriculture and food security is significant and multifaceted. Addressing these challenges will require a coordinated and sustained effort from governments, civil society, and the private sector to promote sustainable agriculture practices, protect vulnerable communities, and build resilience to the impacts of climate change.

Climate Change & Technological Innovations

Technological innovations are playing a critical role in reducing carbon emissions and mitigating the impacts of climate change. There are several innovative solutions that are being developed and implemented across various sectors to reduce carbon emissions and promote sustainability.

One of the most significant technological innovations in recent years is renewable energy, such as solar and wind power. Renewable energy sources are becoming increasingly cost-effective and scalable, providing a viable alternative to fossil fuels. The development of large-scale renewable energy projects, such as wind and solar farms, has significantly contributed to reducing carbon emissions in the energy sector.

Another area of technological innovation is carbon capture and storage (CCS). CCS technology involves capturing carbon dioxide

(CO2) emissions from industrial processes and storing them underground, preventing them from entering the atmosphere. This technology has the potential to significantly reduce carbon emissions from industrial activities, such as cement and steel production.

In addition, the development of electric vehicles (EVs) and other low-emission transportation technologies is reducing carbon emissions from the transportation sector. The widespread adoption of EVs, combined with the development of charging infrastructure, can significantly reduce emissions from passenger vehicles.

Technological innovation is also driving the development of smart grids and energy storage solutions. Smart grids use advanced technologies to manage the distribution and consumption of electricity more efficiently, while energy storage solutions such as batteries and pumped hydro storage can help to balance the variability of renewable energy sources.

Finally, the use of artificial intelligence (AI) and machine learning (ML) is being explored as a way to optimize energy consumption and reduce emissions in various sectors. For example, AI and ML algorithms can optimize building energy management systems, reducing energy waste and improving efficiency.

Technological innovations are providing promising solutions for reducing carbon emissions and mitigating the impacts of climate change. The continued development and adoption of renewable energy, carbon capture and storage, low-emission transportation technologies, smart grids and energy storage solutions, and AI and ML technologies will be critical in achieving a more sustainable and resilient future.

Climate Change Denial & Misinformation

Climate change denial and misinformation refer to the rejection or distortion of scientific evidence indicating that climate change is occurring and is primarily caused by human activities. Climate change denial and misinformation can be harmful as they can undermine efforts to address climate change and delay action to reduce greenhouse gas emissions.

One common form of climate change denial is the rejection of the scientific consensus on climate change. This consensus, based on the work of thousands of scientists around the world, states that climate change is real, and human activities, such as the burning of fossil fuels, are the primary cause of the increase in atmospheric carbon dioxide concentrations and global temperature rise.

Another form of climate change denial is the downplaying or ignoring of the impacts of climate change, such as sea-level rise, ocean acidification, and extreme weather events. Climate change denial can also take the form of questioning the effectiveness of measures aimed at reducing greenhouse gas emissions or suggesting that the costs of addressing climate change outweigh the benefits.

Misinformation can also be spread through social media, where false or misleading information can spread rapidly and reach a wide audience. In some cases, misinformation can be deliberately spread by individuals or organizations with a vested interest in denying or delaying action on climate change.

Climate change denial and misinformation can have significant consequences, as they can hinder efforts to reduce greenhouse gas emissions and limit the impacts of climate change. Delaying action on climate change can result in more severe and costly impacts, such as more frequent and intense extreme weather events, sea-level rise, and food and water scarcity.

To address climate change denial and misinformation, it is essential to promote accurate and science-based information and increase public awareness of the risks and impacts of climate change. Educating individuals and communities on the science of climate change and its impacts can help to counter misinformation and promote informed decision-making.

Climate change denial and misinformation can be harmful and hinder efforts to address climate change. Addressing these challenges requires promoting accurate and science-based information and increasing public awareness of the risks and impacts of climate change.

Climate Change & Public Awareness

Public education and awareness campaigns play a critical role in informing individuals and communities about the impacts of climate change and encouraging them to take action to mitigate its effects. Such campaigns aim to raise awareness, educate the public, and promote behavior change and action.

Education and awareness campaigns can take many forms, including public service announcements, social media campaigns, community events, and educational programs in schools and universities. They can also be targeted at specific audiences, such as policymakers, business leaders, and community groups.

One of the primary goals of these campaigns is to increase public understanding of the causes and impacts of climate change. This can involve providing information on the scientific consensus on climate change, as well as its potential consequences, such as sea-level rise, more frequent and intense heat waves, and more severe weather events.

Education and awareness campaigns can also highlight the actions individuals and communities can take to reduce greenhouse gas emissions and adapt to the impacts of climate change. These actions may include reducing energy consumption, using public transportation or cycling, eating a plant-based diet, and supporting renewable energy development.

In addition to raising awareness and promoting action, education and awareness campaigns can also influence public attitudes towards climate change. By highlighting the importance of addressing climate change, campaigns can help to create a sense of urgency and support for climate action.

Finally, education and awareness campaigns can also encourage policymakers and business leaders to take action on climate change. By highlighting the risks and opportunities associated with climate change, these campaigns can encourage policymakers to enact policies that support the transition to a low-carbon economy and promote sustainable development.

Public education and awareness campaigns play a crucial role in promoting understanding of the impacts of climate change and encouraging action to address its effects. By increasing public awareness, promoting behavior change, and influencing attitudes and policy, these campaigns can help to mitigate the effects of climate change and promote a more sustainable and resilient future.

www.ingramcontent.com/pod-product-compliance
Lightning Source LLC
Chambersburg PA
CBHW071142220526
45467CB00015B/1720